Exploring
Everyday Wonders

Natalie Lunis and Nancy White

Contents

Observing Everyday Wonders

Sparkling dewdrops... flashing fireflies... a shooting star... Did you know that each of nature's everyday wonders holds a science secret? By thinking like a scientist, you can unlock these secrets. Just observe closely, use what you already know, and ask a lot of questions.

Dewdrops on the Grass

Look closely at the grass in this photo. What do you observe? Have you ever seen tiny drops of water on the grass early on a sunny summer morning? Where did the droplets come from?

Science Secret

After sunset, the air close to the ground gets cooler. Moisture in the air forms drops you can see on the grass. The drops are called **dew.** The sun dries up the dew drops the next day.

Think About It

Think about a glass of cold water that you leave out on a hot day. Why do you think drops of water form on the outside of the glass?

Look at this hummingbird hovering beside a flower. How does it look different from other birds you have seen? The hummingbird seems to be floating in the air without moving at all. Where are its wings? Why don't you see them?

Science Secret

The hummingbird's wings beat so quickly that you cannot see them moving. In fact, some hummingbirds' wings can beat more than 4,000 times per minute! The special photo below lets you see the wings.

Think About It

Picture a helicopter on the ground. Now look at this one in the air. How is the helicopter like a hummingbird?

Big Boulders

Have you ever seen a **boulder** in a park, yard, or field? A boulder like the one shown here may be the size of a house and weigh many tons. How do you think this huge rock got here?

Science Secret

Very long ago, giant **glaciers,** or sheets of ice, slowly moved over much of the Earth, carrying along anything in their path.

Think About It

Scientists use facts they already know to make guesses. This kind of educated guess is called a **hypothesis.** Use the science secret to make a hypothesis about how the boulder got here.

Spiderwebs

Why does a spider spin a web? Look at the web in this photo. Do you think it is a home for the spider, or does it have some other purpose?

Science Secret

Spiders produce strands of silk to spin their webs. Some of the strands are sticky, and some are not. The spider takes care not to step on the sticky strands.

Think About It

Review the information in the science secret and the evidence in the photos. Put together the facts to draw a conclusion. Is the web a home or a trap for food?

Reflections in the Water

Put yourself in this scene. Can you see the trees growing by the edge of the pond? What do you notice when you look at the still, clear water?

Science Secret

The smooth, shiny surface of the water bounces light to your eyes just as a mirror does. The image of the trees in the pond is like the **reflection** in the mirror shown below.

Think About It

Predict what would happen on a windy day when there are ripples in the water. How would the reflection of the trees look?

Flickering Fireflies

How would you like to be in this picture, surrounded by the glow of flickering fireflies? What is special about the fireflies? What do you think makes them light up?

Science Secret

Inside its **abdomen,** a firefly has a special chemical that lights up when the firefly wants to attract a mate. An **insect's** abdomen is one of its three main body parts. The other two are the head and the **thorax.**

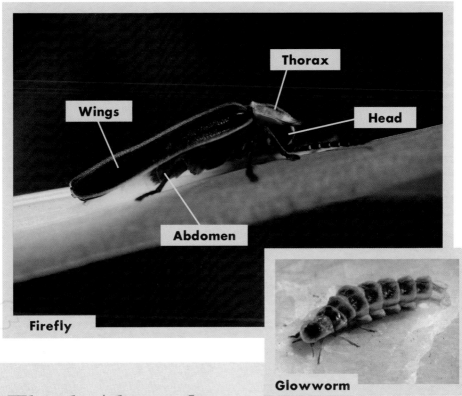

Thorax

Wings

Head

Abdomen

Firefly

Glowworm

Think About It

A firefly starts life as a **glowworm** like this one. How are the glowworm and the firefly different? Compare the firefly at the adult stage and at this earlier stage of its life cycle.

Shooting Stars

Have you ever looked up and seen a streak like this racing across the sky? Many people call this sight a shooting star.

Science Secret

A shooting star is not really a star at all. It is a **meteor**—a streak of light made by a chunk of rocky material burning as it passes through Earth's atmosphere. The photo below shows a chunk of rock that fell to Earth without burning up completely.

Think About It

Suppose your friend thought a shooting star was actually a star—a huge ball of burning gas. How would you explain what a shooting star really is?

Sharing Everyday Wonders

Once you start exploring the everyday wonders in nature, you can make some very interesting discoveries. Tell your friends what you have learned. After all, part of being a scientist is communicating your ideas to others.

Glossary

abdomen (AB-duh-mun): the rear section of an insect's body

boulder (BOL-dur): a large rock

dew (DOO): droplets of water that form on surfaces when air cools at night

glacier (GLAY-shur): a giant sheet of ice that moves slowly over Earth's surface

glowworm (GLOH-wurm): a firefly at an earlier stage of its life cycle

hypothesis (hye-PAH-thuh-sus): a guess based on known facts and scientific thinking

insect (IN-sekt): a small animal with three main body parts, three pairs of legs, and sometimes one or two pairs of wings

meteor (MEE-tee-ur): a streak of light made by a chunk of rock or metal, burning as it speeds through Earth's atmosphere. The piece of rock or metal comes from space.

predict (prih-DIKT): make a careful guess about what will happen. When scientists predict, they use what they already know and what they observe.

reflection (rih-FLEK-shun): a picture formed by light bouncing off a smooth surface

thorax (THOR-aks): the middle section of an insect's body

Index